NUMBER 570

THE ENGLISH EXPERIENCE

ITS RECORD IN EARLY PRINTED BOOKS
PUBLISHED IN FACSIMILE

ANDREW BORDE

THE PRYNCYPLES OF ASTRONAMYE

LONDON, (AB. 1542?)

DA CAPO PRESS
THEATRVM ORBIS TERRARVM LTD.
AMSTERDAM 1973 NEW YORK

The publishers acknowledge their gratitude to
the Syndics of Cambridge University Library
for their permission to reproduce the
Library's copy, Shelfmark: Syn.8.54.82

S.T.C.No. 3386

Collation: [ ]⁸, B-D⁸

Published in 1973 by

Theatrum Orbis Terrarum Ltd.,
O.Z. Voorburgwal 85, Amsterdam

&

Da Capo Press Inc.
a subsidiary of Plenum Publishing Corporation
277 West 17th Street, New York N.Y. 1011

Printed in the Netherlands
ISBN 90 221 0570 9

Library of Congress Catalog Card Number:
73-6101

The pryncyples of Astronamye the whiche

diligently perscrutyd is in maner a pronosticacyon to the worldes end compplyd by Andrew Boorde of physick Doctor

The preface

It is gretly to be dolentyd ẏ all men almoſt dothe nowe adnichilat the .vii. lyberall ſcyences ſpecyally, Aſtronamye the whyche of trewth is a neceſſary ſcience And doth illucidat all other ſcyences, not only ẏ liberall ſcyences as grāer retoryck logick Arithmetrick geometrick and muſick, but alſo hit is an Introduccyon to philoſophy, phyſicke ⁊ diuinite it is not vnknowen ẏ philoſophy conſyſceyth in naturall phyloſophy ⁊ morall philoſophi, to ẏ naturalite ⁊ moralite of the which no man can attynge w̄out Aſtronamy, Alſo Aſtronamy doth illucidat phiſicke for hypocrates ſayth, Cecus medycus eſt qui, Aſtronamyam neſcyt ẏ is to ſay he is a blynd phyſycyon ẏ whiche doth not know Aſtronamie alſo Aſtronamye doth geue knowlege to dyuynyte as hyt doth playnly, appere in the fyrſte Chapter of ẏ geneſys, and dyuerce other places

places Wherfor dauid doth say Celi enar rant gloriam dei et opera manium eius annunciat firmamentum. heuyns doth shew þ glory of god & the fyrmamēt doth shew the warkis of god I consyderyng thes prempsses with the gret, helpe of the afforesayd s. pryer to mak this boke to anymer al men to haue a respect to hit and to study hit to laude & to playse god in his warkis and Whcr I haue omytted & lefft out many matters aptaryng to this boke late them loke in a book naympd the Introduction of knowleg a boke of my makyng the Which ys apzintyng at old Robert Coplands the eldyst printer of In gland the Which doth print thes pyre mi pronosticacions

Finis,

¶ The Capytles of conternces
of thys boock folowth

The fyrst Capytle doth shew the names of the .xii. sygnes and of the .vii. planetes. And what the zodyack and how many minutes a degre doth cōtaine

¶ The seconde Capytle doth shew what sygnes be mouable, and what sygnes be not mouable and which be commone and which be masculyn signes and which be femynyne and of the trypleycyte of them.

¶ The .iii. capytle dothe shewe in what members or places in mā ÿ sygnes hath theyr domynion and how no man owt to be let blod whan the moone is in ÿ sygne wher the sygne hath domynyon and also what operacion the sygnes be of whan ÿ moone is in ther.

¶ The .iiii. capytle doth shew of the fortitudes of the planetes and what influens they doth geue to vs.

The

¶ The.v.Capitle doth shew the natural dyspocycyon of the mone whan she is in any of the.xii.sygnes.

¶ The.vi.capytle doth shew of ẏ nature of al ẏ.xii.sygnes And what influēce thei hath in mā. And what fortitudes ẏ planetes hath in ẏ sygnes. w̄ the names of the Aspects.

¶ The.vii,capytle doth shew ẏ natural dyspocycions of the planetes. And what operacyon they hath in mans body.

¶ The.viii.Capitle doth shew of the .v Aspectus .and of theyr operacyon

¶ The ix capitle doth shew of ẏ mutaciō of ẏ Ayer whan any rayne wind wedder froste and cold, shold be by the course of ẏ sygnes and planetes.

¶ The.x.capytle,doth shew ẏ pedyciall of the aspectus of the mone and other planets and what dayes be good .and what dayes be not.&c.

¶ The.xi capptle doth shew of flenbothomy or lettyng of bold

¶ The xii capitle doth shew how whan & what tyme a phisicion sholde minister

medycynes

¶ The .xiii. Capitle doth shew of sowig of seedes & plantynge of trees and setyng of herbe.

Thus endyth the table.

¶ The fyrst capytle of the ptract of ỹ names of planetes and also of the names of
the sygnes and what the zodiack
is and how many degrees be
in ỹ zodiack & how many
minutes a degre doth
contayne.

The names of the .vii. planetes be
thes. Saturne Jubiter. Mars ỹ
Sone. venus Mercury and the Moone.
¶ Thes be the names of the .xii. sygnes
Aries. Taurus, Gemini. Cauncer Leo.
Virgo. Libra. Scorppo. Sagittary. Capricorne. Aquary. & Pisces. The zodiacke
is the Circle in the which the .xii. signes
be in namyd the circle of beestes. or ỹ ob-
lique cyrcle hit may be called wel ỹ circle
of bestes for the sygnes within the cyrcle
hath or be of that dyspocycyõ of ỹ bestes
of whome they doth take theyr names.
or be lykyned vnto. As Aries is lykinned
to the Wedder or Rame. Taurus is lyke
to the Bull. Geminy ys lykyned to
tow

.ii. chyldzyne bozne at a bozdyn. Cancer is lykened to the Crab oz Craups. Lio is lykned to the Lyon. Uirgo is likened to a Mayd. Libza is lykened to a payze of balaunce. Scozpio is lykened to a venemus wozme namyd a Scozpion. Sagettarius is lykened to a monsterus archer oz shoter the which is lyke a man frō the mydle vpward and lyck a foze foted best from the mydle down ward. Capzycoznus is lykened to the watter Pisces is lykened to fyshes Thes sygnes geueth in fluence to vs by leke operacyon as the bestes be of nature. Euery one of thes sygnes is deuydyd in .lx. degres And euery degre doth contayn .vi. mynutes. So that in ƶodiack is .CCC.vi. degres.

⁋ The seconde Capitle doth ptract of the sygnes which be mouable And which be not mouable. And which be comone And which be maculyn signes, and which be femunyn And of the triplycyte of them

Thes sygnes be mouable. Aries Caucer. libra. and Capricorne. And thes signes be fixed. Taurus leo Scorpyo and Aquary. Thes sygnes be comne Gemini. virgo. Sagittary and pisces ⁋ Thes signes be masculin signes, Aries. Gemyny. leo libra. sagittarius and aquary And thes be femynie signes. Taurus. Caucer vyrgo. Scorpyo capzycorne and pysces. ⁋ Also ther be. iiii. triplycytes. Ozientall meridronal Occideutal And ẏ Septrionall The orpent trypliclte the whiche is in the est is colorycke fyre and masculyne And this triplicite hath thers. iii. sygnes Aries. leo and Sagittary. The lordes or planetes of this triplicite is the sone. in ẏ day. And Iubiter in night And Saturne is in both. And. this triplicite is dyurnall

The second tryplicyte is the meridionall triplicyte which is in ẏ sowch. And this tryplicite hath other .iii. signes. Taurus virgo an Capricorne This tryplycite is melancoly & feminyne The lordes or planetes of thys tryplpcyte is venus in the day And the mone in ẏ night And mares in both And this triplycyte is nocturnal The therde tryplpcyte is the occidentall tryplycyte. whiche is in the west And is sanguyn & femynyne And this triplycite hath other .iii. signes. Gemyni lybra and Aquary. The lordes or planetes of thys trypliycte is Saturne in ẏ day marcury in the nyght And Jubyter in both, And thys triplicite is dyurnall The .iiii. triplycyte is the Septemtryonall triplycite which is flewmatick & femynyne & this trypliycite hath .iii. other sygnes. Cancer Scorpio & pisces. The lordes or planetes of this triplicite is venus in ẏ day mares in the nyght the Moone in bothe.

¶ The .iii. Capytle doth shew in what members or places in mā ẏ sygnes hath
                                                   theyr

ther domynyon or power And
how no man owght to
be lett blod whan the
moone is in the
sygne wher
the sygne
hath
domynyon And also what opercion
the sognes be of whan the
moone is in them

The .xii. sygnes hath domynyon
vppon the members of every
man, for Aries beholdeth the
hed. Taurus the necke. Gemini the armes & handes Cancer
the brest Leo the hart. Uirgo ye bely. Libra the raynes & buttokes. Scorpio the
secret places of man Sagittary the thyes
Capricorne the knes Aquary the legges
pisces the feet. Let every man be ware
yt not flebothomator or letter of bloo. nor
no maner of Chirurgyon do tuch hym in
opnyng any vayne or do mak any incicion
or cutting whan the moone is in any signe
wher the sygne hath any   domynyon

dominion oz deth regne also whan ẏ mo=
one ys in a mouable senge hit ys good to
tak a voiner and good to tak a Jurney oz
to goo about any maner of busynes oz to
begin any matter pertayning to worldly
police whan the moone is in a fexed signe
hit is good to mak fundacions of howses
oz walles good to bild ɟ to redifi ɟ to goo
about stable warkes whan the moone is
In a comune signe hit is not good to mak
no mareẏ noz to mari noz to mak no barg
yn with out record and sure writting noz
good foz sick men ɟ significth much in sta
bilite.

¶The iiii capitle doth shew of ẏ
fortitudes of planetes hauing
oz geueng theyr
influence
to vs

The planetes hath .v. fortitudes oz
streaghtes oz testimonialls which
be thes ẏ howse ẏ exaltacio̅ ẏ Triplicyte
¶The terme And the   face.

face. The howse hath .v. fortitudes, The exaltacyō hath .iiii. fortytudes. The triplicite hath iii fortitudes. The terme hath ii. fortytudes and ye face hath one ¶ Euery planete in his howse is as a kinge sytteyng in is mageste in a parlament. Euery planet in his exaltacyon is lek a kyng that is a crownyng euery planete in his replicyte ys as a kyng in actoryte amongs his lordes which be his helpers in euere planete in his terme ys as a man that ys a mongs his freds and kings folke the face geueth to a planete roome or space as a chayer or a set is for a master

¶ The v capytle doth shew the
naturall dyspocycyon of
the moone whan she
is in any of the
.xii sygnes.

When the moone is in Aricte. hit is not good for no syche man nor sickely men to shaue theyr hed or berd. for euery here hath a hoole. by the which euyll vapoures be in haustyd. Also hit is not good to medyl w the eres nor eyes nor tong nor no other place in or abute the hed w no instrument of Iren nor to lett blod or to boxe or cupp any place or vayne about the hed and necke & be ware that no blod be exhaustyd owt of a va----

namyd Cephalica. But it is good for ba=
thes. and also it is good to begin any thig
that wolde haue short expedycion & good
to goo about ani matter that a man hath
to do with gret men or rych men

℄ Whã the moone is in tauro hit is not
good to medyll with the necke to exhaust
any blod But this time is fortunat to ma=
ny thynges. it is good to bay oxen kyne &
calues hit is good to byld any howse or
mansyon or to begyn any other thynge it
is good to plant & to Impe or to graf eres
it is good to sowe al maner of corne or see=
des and good to comon with women &c.

℄ Whan the mone is in gemini exhaust
no blodo wt of the armes nor hãdes it is
good to putt children to skole or to any sci=
ence or crawght and good for a man to be=
gyn that thing that he wold haue it reite=
ratyd or begon agene.

℄ Whan the moone is in cancro a man
may be lett blod without parell and ptho=
lomp sayth it is a good tyme to take med=
sons specyally electuares also it is good
to tak

to tak any iorney and to remoue from one place to an other and from one castyll or mansyon to another

¶ Whan the moone is in Leone it is good not to tak any vomyte for hit wyll hurt ẏ hart and brest & it is good to begyn that thynge that a man wolde haue continew

¶ Whan the mone is in Virgine it is not good to mak any incycyons for ruptures nother to medle with the huckyl boones, to exhauſt any blod, to mary in this time is not praysed leſt ẏ woman bere no frute to obtayne the loue of a mayd in the way of marage is good and Also this tyme is good to sett chyldren to skole.

Whan the moone is in Libra exhauſt no blod from the buttockes & tak no medsons frome the raynes of the bake thys tyme is good for a man to bay or to sell & good to take a iurney & thys tyme refrayne from all veneryouse actes also this tyme is good to go about ani matter bisines workes or of doynges

Whan the moone is in Scorpione be
ware

ware of al maner of incysion for the stone
for harnyes & for ruptures. And be ware
of bathes this tyme is good for few thyn
ges or noone.

☙ Whan the moone is in Sagittario.
exhaust no blod owt of the thyes to haue
busynes wt iudges iustices & men of law
it is good & this tyme is good for Archers
And marchautes & good to mak bargins

⁋ Whā ẏ mone is in Capricorne exhaust
no blod owt or from the knes this is an
euyll tyme to take any medycynes hit is
good to ere and to plante & to sett herbes

☙ Whan the moone is in Aquari hit is
not good to exhaust any blod out of ẏ leg
ges nor to take no mediceus. hit is a beri
good tyme to byld

⁋ Whan ẏ mone is in piscibus. be ware
of exhaustinge of any blod owt of the fete
and legges nor medyl with no sicknes or
dyseses in the fete or legges this tyme is
good for them that wyll take any iurney
and goode for Feshers and for them that
vsyth the water.

                              The.

¶ The .vi. capitle doth shewe of the natur of all the sygnes. and what influence they hath in man. And what fortitudes the planetes hath in the sygnes. with the aspectes.

Aries doth gouerne the hed. and is a masculine signe. and is hote and hath his dominion in the howse of veith and lyef. geuynge influence to them the which be borne vnder him to haue a longe neck. a lene face. short eres and somwhat hery. Taurus doth gouerne the neck and is a feminin signe and is colde & drye leke to melancoly. And is in the howse of substaunce and possession. of gyft and receite geuing influence to the y be borne vnder this signe to haue a brode face. a gret nose grete eres gret nostrells and a gret necke Saturne in Tauro hath. ii. fortitudes. ye terme & the face. Iubiter in Tauro hath one y is y terme. Mars in Tauro hath. ii

fortitudes. A triplicite and a terme. The
sone hath none. Uenus hath. iii. ẏ howse
the triplicite & the terme. Marcury hath
.ii. The terme and the face. The moone
hath. iii. fortitudes in Tauro exaltaciont
the face. And the triplicite. Gemini dothe
gouerne the Armes and handes. And is
a masculyn syne and is hott & moyst, leik
the Ayer and lyke sanguyne complexion.
And is in the howse of kynred fraternite
and counsell, geuyng influence to them ẏ
whiche be borne vnder this sygne to be
bewtyfull and fayer and comly. and som
nolent. Saturne in Gemini hath. ii. forti
tudes. A terme. and a triplicite. Jubiter
hath. iii. A triplicite. The face and terme
Mars hath. ii. The face and terme. The
sone hath one which is the face. & Uenus
hath one which is The terme. Marcury
hath. iii. which be thes. The triplicite ẏ
howse and the terme and the Mone doth
not posses no dyngnite in gemini. Cancer
doth gouern the Brest the longes and the
Somacke. And is a feminine sygne. And

ys

is colde and moyst lyke the watter and ẏ
complexion of flewmatyck persones And
is in the howse of the father of hosbonds
and wyues. Cytes. Tresures. burpalles
and herytag. this sygne is also vnstable.
Wyndy and waitery she geuynge influen=
ce to them the which be borne vnder this
sygne be grose byneth and slender a boue
¶ Saturne in cancro hath but one forti=
tude which is. The terme Iubyter hath
.ii. fortitudes. The exaltacyon & the terme
Mars hath. ii. a triplicite and a terme.
The Soone hath noone Uenus hath .iii
A triplicite. a terme & The face Marcuri
hath. ii. The terme and the face and the
Moone hath. iii. the howse the triplicite
and the face.

¶ Leo doth gouern the harte and ẏ lyuer
And is a masculine syne. and is hote and
dzye lyke the complexion of colerike men
And is in the howse of honor. and childzen
And doth geue influence to them the whi
che is borne vnder this sygne to be grose
aboue

aboue and slender vnder neth the waste of Saturne in Leo hath. iii. fortitudes. ÿ triplicite the face and the terme Jupiter hath. iii. The. Triplicite the terme and ÿ face. Mars hath. ii. the terme and ÿ face The Sone hath. ii. ÿ howse & Triplicite Uenus hath one which is the terme And the moone hath noone.

Uirgo doth gouerne ÿ intrayles or wombe And is a feminin sygne And is colde & dry leek the erth And complexion of melācoly parsons And is in the howse of sycknes. And cattell And doth geue influence to the ÿ which be borne vnder this signe to be beutyfull haupnge a fayer & louyng bisage ¶ Saturne in Uirgine hath one fortitude namid the terme, Jubiter hath one nampd the terme Mars hath. ii The Triplicite and the terme. The sone hath one which is ÿ face, Uenus. iii. The triplicite the face, and the terme Marcury ha the fiii. the howse. the terme. ÿ exaltacion and the face. The Mone hath one which is a Triplicite

Libra

⁌ Libra doth gouerne the raynes of the backe and ẏ buttockes and is a masculin signe and is hote and moyst lyck the ayre And sanguine psons. And is in ẏ howses of mariage theffcth. strife and robery. and doth geue influence to them the which be borne vnder this sygne to euyll dispocicions onles grace wark aboue nature Saturne in Libra hath .iiii. fortitudes. The exaltacyō ẏ terme ẏ Triplicite and the terme and the face Mars hath one which is the terme. The Sone hath none. Uenus hath .ii. The howse and the terme. Mercury hath .ii the Triplicite and the terme And the Mone hath one. Which is ẏ face

Scorpio doth gouerne ẏ secret members of man and woman. And is a femynyne signe. And is cold and moyste. lick ẏ water ⁊ the complexiō of flewmatick psons And is in the howse of deth sere.

debate

debate. warre. oet labor chastyte and wit And doth geue influence to them the whiche be borne vnder thys sygne to haue a ruddy face slender and smale leges. And grete fete ⁋ Saturne in Scorpione hath one fortitud which is the terme Jubiter hath one which is the terme. Mares hath .iii. which be thes ẏ howse, ẏ terme and the triplicite. The sone hath one whiche is the face Uenus hath .iii. The triplicite the terme & the face. Marcury hath one which is the terme. The Mone hath one which is the triplicite.

⁋ Sagittarius doth gouerne ẏ flanckes And the thyes And is a masculyne sygne And is hotte & drie lick the fyre & colorick psons. And is in ẏ howse of honour. And master shepe by way of offyce. And doth geue influence to them ẏ which be borne vnder this signe to be wyse and to haue knowlege in. Astronami. And shold haue a longe bisage and a bygge bely and not gretly hered but suffycyent and strayght

Saturne

Saturne in Sagittario hath .iii. fortitudes. The Triplicite The terme and the face Jubiter hath .iii. the howse the Triplicite and the terme Mars hath one which is þ terme The sone hath one whiche is þ Triplicite Venus hath one whiche is the terme. Marcury hath .ii. þ terme and the face. The mone hath one whiche is the face.

¶ Capricorne doth gouerne þ knes. And is a feminye sygne, And is cold & drye as the erth, and melancoly psons and is in þ howse of lordshep, honour, and substance And doth geue influence to them the whiche be borne vnder this signes to be slender legged, hauynge a drye body & a hery face, repletyd wt melancolines
¶ Saturne in Capricorne hath .ii. fortitudes, the howse and the terme. Jubiter hath .ii. the terme and the face. Mars hath .iiii. the exaltacion þ terme þ Triplicite and the face. The sone hath one which is the face. Venus hath the .Triplicite and
The.

the terme .Marcury hath one which is ẏ
terme. And the moone hath one which is
ẏ triplicite Aquarius doth gouerne ẏ leg﹑
ges bi neth ẏ knes to ẏ feete. it is a mascu
line signe And is hote and myeste as the
Aier lyek to the cõplexeon of sanguine mē
And is in the howse of frendship and sub﹑
stance and doth geue influence to hem the
Which be borne vnder this signe to be a﹑
miable how be hit the one lege most cõme
li is longer than an other or els bigger th﹑
an an other and such psons shold be a boſt
er and a geet waster ☾ Saturne in a qua
rio hath .iii. fortitudes The howse the
terme and the triplicite Jubeter hath .ii.
the triplicite and the terme, Mars hath
one which is the terme, The son hath no
one. venus hath .ii. the Terme and ẏ face
Marcury hath .iii. The triplicite the ter
me ƶ ẏ face The moone hath one which
is the face

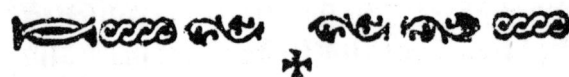

¶Pisces doth gouerne the feete. And is a femynyn signe And is cold and moyste as the watter liek to fleumatick psons and is in the howse of heuynes and enemyes and rydinge bestes. And presonment. And doth geue influens to them the which be borne vnder this signe to be of a larg brest a litle hred and whitly faced, rownd eres & of good corag ¶Saturne in piscibus hath .ii. for iudes The terme and the face Jubitor hath .iii. the triplicite the terme and the face The son hath noone, venus hath .iii. the exaltacion the terme and ye triplicite Marcury hath one which is ye terme and the sone hath one whit he is ye triplicite.

Now her is to be notid for Aspectes wich be thes .A. sextyl a terme. A quartyple a coniuccyon and an opposicyon Euer mor hauenq a respect to ye Aspectes of the planetes. And yf the

Aspectes

Aspectes be good of ẏ mone. & the signes aboue nampd be good. the matter shalbe of affecte and if the aspectrs be euyll of ẏ Moone the matter will take lytle affecte or none. And which be good aspectes and which be euyl. shalbe shewed in the next capytle after this that foloweth.

❧ The .vii. capitle doth shew of ẏ nature of the planetes. and of theyr operacyon in man.

As necessary hit is to know ẏ nature of the planetes with ther operacione. as it is to know the nature and operacion of the sygnes. Saturne is the hyghest planete which is colde and drye lyke ẏ erth hauyng a gret interest in malancoli men. This planet is a masculine planete And his howse in Capricorne. & this planete hath an interest in the clargi or spirituallte for he enducithe dingnite pontificalnes polyce and ingeniousnes.

how be it he is a signifier of such as doth
warke in vile handy chraftes. ⁊ is enimi
to ẏ lyf of man for he is enemy to nature
And is a slow planete in fetchynge his
cercuite or curcunference ⁊ they ẏ which
be borne vnder hym be slow ⁊ not quick
hasty. ⁊ malycyouse except grace warke
aboue nature Saturne doth gouern i mã
Melancoly. the splen. the gal the bladder
And hath a pticipacion of flume And he
doth ingender thes infirmites. splenyt sic
knes and pticis and the fener etick be cõ
sumpcyons. Also he is one of the causers
of the .iiii. kyndes of leprosnes. The .iiii.
kyndes of the poropises The .iiii. kindes
of the gowtes. The to kyndes of the pal
syes ⁊ Roume descendyng frome the hed
to the brest namyd catarrus. And ẏ Can
ker and all maner of sicknes in the whi
che is any coldnes for this matter looke
in the Breuyary of helth and in the In
troduccyon of knowleg

       Iubyter

¶ Iubiter is the second hyghest planete
which is hoot and moyst lick to the Ayer
and he is a masculyne planete hauynge
an interest in sanguine persons. and hys
howses be in Sagittari and Pisces and
he is lorde ouer the temporalte. Specialy
of chyualry. and men of estate. as a beni-
uoluse planete. how behit he is socwhat
slow in fetchyng his circuite. and they ÿ
which be borne vnder this planete. shold
be manly. comly. gentyll and honest. &c.
Iubiter doth gouerne in man the longes
the nature of man and womā. the boones
with the cartalages. and thes be his sicknesses. The squince. periplum ōia which
is an impostume about the longes. the
plurese the Crampe The Appoplexe. the
Litarge the Cephalick and the Cardiack
and such lyek the which doth com by the
occasion of blode.

¶ Mars is the therde planete. Whiche
is hoote and drye leke to the fyre. And
is a Masculyne planete. And hath interest

reſt in colerick perſons. and his howſes
be in Aries and Scorpion. and he is full
of malyce wrath and vengyons. furious
and full of bate and ſtryfe. how be hit he
is indyfferent in fetching his circumference.
and they the whiche be borne under
this planet except grace warke aboue na
ture ſholde be a lyer a robber a morderer
furyouſe and vengable. ful of wrath and
angry haſty and teſty and euell diſpoſed
And he doth gouerne in man & woman ÿ
vaines and the teſticles of man and coler
And albumeſer ſaith he doth gouerne ÿ
Lyuer. And I ſay he hath intereſt in the
gall of a man & woman. And thes be his
ſickneſſes the feuer tercyane ÿ continewi
an & the cauſen and al other hot feuers. &
hot appeſtemacions. Alſo the migrime.

And woman that be delyuerid befor theyr
tyme and all other infirmytes the which
doth com of extremee hete. The ſone is in
the mydle of the .vii. planetes. geuynge
as muche lyght vpward to heuen as he
doth

doth downward to vs The son is hotte & drie lyek ẏ fire & yet he is a frend to blod & singuin persons And his howse is in the Lyon he is a masculin planete beniuolense and good in all thyngs and a confortor of all terrestiall thyngs And doth induce naturall strength and is a frend to gret men of Honor And he doth fetche his cercuite ons in a yere. And they the which be borne vnder the sone shold be strong Welfanerd & lusti wise descret and riche, and he doth gouerne in mā and womā the hart. the brayne. the mari with in the boones the senowes and the syght. how be hit thorow his vehement hete in somer he doth hurt the hed & doth calist or chastith ẏ blod to much of ẏ which he hath vigor and strenght,

⸿ Venus is a pulcrus planet. and is cold and moyst lycke the watter and hath interest in blod and fleme she is a feminine planete and her howses be in Tauro and Libra. she is lady ouer loupers. and she doth fetche hyr cercuite almost as soone
as

as the Sonne She doth induce loue and
veneryousnes. And he or she that is borne
vnder this planete shold be a true louer.
Louynge. Merthe melody and musicke
louyng also syngyng and dawnsyng. In
terludes games & sportes and gestes and
they shold be fayer and amyable & full of
bewty and plesant. And She doth gouer
ne in man & woman the seed of man the
rygbones the loynes the matrix and ye
secret partes. and the farnes in man and
woman and ye smellyng. and thes be hyr
sicknesses. coldnes of the stomacke ye slac
knes of the lyuer to dygest. the passyons
of the hart the Iliaca passio the colyk the
fystyls the plyes ye hemerodes ye suffoca
cyon of the matryx & other infyrmytes y
be about the secret places of mā & womā
ye commyth of superflug humores. Mar
cury. is temperatli hote & dry. good with
goodnes and euyll with euylnes. he is a
masculyne planete and his howses be in
Gemini and virgo and he doth fetche his
cercupte almost as soune as the sonne. He
        C.i.   doth

doth induce ingenyosues & gret knowleg of dyuers artes, facultes & seyences. And good in trauel and messages doyng. Who so that is borne vnder this planete sholde be gentyl & amabell And a gret traueller and good to send or goo in messages and he sholde expert in sciences. for he sholde haue a good wytt & cast fare in polytycke reson. Marcury doth gouerne ÿ memory and the toung and the gal. And thes be his enfyrmytes. the .iii. kyndes of madnes. ÿ iii. kyndes of the fallynge syckues. The cowgh ÿ percipitacion of the matrix. and parturbacio͂ of mynd & other infirmites the which co͂meth of siccite or dryues.

The Moone is the lowest planete. And is cold and moyst lyck the watter, and hath an interest in flumatyck parsons. and she is a feminine planete and is mother and a ministrix to all moyster and is lady and gubernatrix ouer the seeis and watters and her howse is in Cancer and she doth fetch her cercuite ons in a mounth. And she doth induce mutabilite, inconstance.

Trauell

Trauell, message ambassage. And fortunate to aquatical maters or bysines, they that be borne vnder the Mone sholde not be stabyll wytted, and shold be a trauelet and seke strang coūtres. The Mone doth gouerne in man or woman the stomacke and the bely and the secret place of womē and hath also an interest in the hed. And thes be her infirmites, comocyon trymblyng or shakynge of members vniuersall or pticuler the cardyack the palses & such leke. ☞Who so euer wolde haue a forder noticion of the sygnes and the planetes & of theyr influence or consteliacyon, let thē look in a book that is now a pryntyng at Robert cooplondes namyd the Introduccyon of knowleg. And he that wyll haue the knowleg of, all maner of sicknesses & dyselps let them looke in the breupary of helth whiche is pryntyd at Wyllyam Mydyltons in flet stret.

☞The

¶ The .vii. Cappyle doth shew of the
.v. aspectes of the planetes.

S I sayde in the .vi. cappyle
ther be .v. Aspectes of the planetes. A coniunction, A sextile,
a trine a quartile and an opposicyon. A coniunctione pperly
is not takyn for an Aspect, for an aspect
is a dystance of signis. But planetes coniunct, doth not differ by distance of degrees, for a coniuctio amonges y planetes
is when the planetes doth draw to gyd-
der by lesser dystance than .xii. degrees. &
that one of them failith vnder the bright
beemes of an other, than it is namyd coi-
iunct to the sonne. And when the planete
doth drawe nygh to y sone by .xv. degres
than is y planete combustid by y splendre
or beemes of the sone. And than dyuerce
times happynith rayne as it doth, when
the sone and the Moone be almost to gy-
der in ther orbs or cercles, than in the end
of that moone shold be rayne This aspect
which

which is a coniunction And firſt in order
is namyd a beniuolēt or a frendful aspect
for good to goodnes is incresyd of a gret
and hygh beniuolence and vertu. And yf
goodnes be Joyned to euil, thā ÿ euilnes
takyth a way half the goodnes. wherfor
he that wyl go about to do any good thig
owt to mark the mone specyally yf ſhe be
coniunct to good planetes or els ſeperatid
from eupll planetes And he that wyl go.
a bout an euil or a ſhrod turne muſt choſe
the tyme whan the moone is coniunct w
eupl planetes, or els ſeperatid from good
planetes.

⁋ The ſecond aspect is a sextyle, that is
whan the planetes be dyſtant one frome
another by. vi. partes of the zodiack. as
thus in makynge a ſimilitud of, ii. planetes, that venus be put in the begynnyng
of Aries, and another planet in ÿ begyn
nyng of Gemini Or els one in the mydle
of Aries an) another in the the mydle of
Gemini and ſo in leke maner of all other
planetes

Planetes . but this . Aspect is not pfyt in frendshep for the aspect is not certin mor oppn but pryne

⁋ The therd Aspect is a quartyll. And y is whan a planete hath an . Aspect to another planete. by dystaunce of .iiii. sygnes the which dystaunce is the .iiii. part of the zodyack. as thus yf one planete be in the begynnynge of . Aries and another in the beginning of Cancer. Or els one planete in the mydle or end of Aries. and another in y mydle or end of Cancer. Such aspect is inpfit of hatred enuy and traytery.

⁋ The .iiii. Aspect is a tryne. And that is whan one planete is dystant frome another by .iiii. sygnes. the which distance is the therd part of the zodiack as thus, yf one planete . ware in the begynyng of Aries. and the other in the begynwyng of Leo this is a manifest and a perfit aspect of frendshep

⁋ The .v. Aspect is an Opposicion & that is whan one planete is distant from another by half of y whole zodiack as thus
that

that one planete be in the begynnynge o:
in the midle o: end of Aries and the other
be in ẏ begynnyng mydle o: end of Libra
takyng both lyek and so is hit in al other
sygnes This aspect is an euident & mani
fest aspect of hatrede of warre and other
displesures

☙ The .ir. capitle doth shew of ẏ mutaciō
of the yper whan any rayne, wynd thon
der haple frost o: snow shall fall.

Ho so euer ÿ wyll pnosticat of the mutacion of the Aier he must fyrst mack the .iiii. tymes of the yere. Whiche be to sai ver estas Autûnus & hiems that is to say. the spring ÿ somer harupst. and wynter. Uer which is the sopyng by nature shold be hote and moyst And in this tyme blod doth moue and encresyth w.th all other naturall thynges, Estas Whiche is somer is hote and dry & than coler doth moue & incresith. Atûnus o: harupst is cold and dry & hath respect to melancoly, hyems Which is Wynter is cold & moyst & hathe respect to slume. Secundaryly. one must mark the nature and the qualytes of the sygnes, Which be moyst Which be colt e And which be drye Therdly one must consider the nature or pyertes of the planetes. for the howse of coniuracyon is Jubyter. The howse of tempestes is Mars, & yf they be coniûct or be in a quartyle aspect both Wil happê Yf. saturne & venus be coniunct, or be in a sextyle

a certyle than hit wylbe cold, Jubyter is
the sone.or in an opposytine aspecte wyl
haue grett wynd Mercury with venus
causeth raine ¶ Forder more a coniucty
on of Saturne with the mone.maki h mu
tacyon in the Ayer for rayne & grose clou
des be elenatyd with cold.specially wha
this coniunctyon is in any aquatycal or
terrestyall sygne. ¶ A coniunctyo of Iu
byter with the moone causeth dyuerce &
many whyt cloudes. ¶ A coiunctyo of
Mars with the moone in moyst sygnes
casyth rayne ¶ A coniunctyon of the son
with the mone.causyth rayne specyally
in moist signes. ¶ A coniunctio of Venus
w the mone in moistsignes causeth swet
dwes and soft rayne ¶ A coniunctnon of
the moone to Marcury in moyst sygnes
causyth rayne w wynd. ¶ A coniunction
of saturne with Jubiter.iii.daies before
or.iii.dayes after causeth gret mutacyo
in the Ayer. ¶ A coniunctyon of saturne
with mars in moyst sygnes.iii dayes be
fore and after the Ayer to be co uptyd by
the which

the whch many men women & chyldryn be infectyd with dyuerce disesis specialli with the pestylence. And dyuerce tymes causyth hayle rayne lyghtnynge & thondryng. this coniunctyon for mā or womā is moyst worst ¶ A coniunctiō of saturne with the sone. causith cold rainy wedder ¶ A coniunction of saturne with. Uenus coniunct in eny aquatycal signe makyth cold raynes to indure ¶ A coniunction of saturne w Marcury in aquatycal signes causyth rayne And in dry sygnes causyth, drynes, ¶ A coniunctyon of Mars w marcury. in hote sygne causyth. Hete in dry sygnes. in aquatycall sygnes rayne.

¶ The.x. capytle doth shew the
Judiciallnes of the Aspectes
of the Mone another
planetes what
dayes be
good.
.K.

A Sextile aspect of the Moone to Saturne is a good day to sped matters with women & to take counsell of lernyd men or wyse mē, this day is good to begyn byldyng or to rehedyfy. good to to tyll or dyg the growne and to sett and plant herbes and tres.

¶ A sextile aspect of ẏ Mone to Jubiter is a good day for a man to spek to a spirituall Judge & to lawyers. this day after astronamy is good for all matters.

¶ A sextile aspect of Mars to the Mone is a good day to speke to any gret man of noblylyte. and good to comone or talke w strenuose & armigernse men of berth. and good to bay any thyng. And to sett forth horses w horsemen in war seson.

¶ A sextile aspect of Uenus to the Mone is a good day to mak a maryage. and to seke loue and fauer of any parson & good to speke to any noble parson.   A sextil aspect of marcuri to ẏ mone is a good day for a mā to mak a cōpte, & to shew or plet his caues & good for a mā to by or sel or to

mak marchauntes. And good to set chyl
dryn to lernnyng And good to take a Jur
ney. and good to make meter or orayons
or to prech. or rede a lecture.

℗ A sextyle aspect of ẏ sone to the moone
is a good day to mete & to salute kynges
and princes and to aske or disyer lefull &
lawfull thynges & a good day to sek owt
ẏ trewth of a matter. & good to assocyatt
wich wyse men

℗ A tryne aspect of saturne to the moone
is good to be assocyat or in the company of
old men or womē and good to dydg delue
or ere the growne or to sowe any seedes.
good to plant and to set herbes and good
to make fundacyons to byld

℗ A trine aspect of Jubetor to the moone
this is a fortunat daye for all maner of
matters or causes and very good to spek
to iudges and men of lawe specyally of ẏ
spiritualte

℗ A tryne aspect of. Mars to the moone
to be with the company of knyghtes and
squyers and suche lick. And to

pcure

pcure ordin g spare for all thyngs prey=
yng to wat and good to bay bestes

⁋ A trine aspect of the sone to the moone
is a good day to aske a pdon of a kyng &
good to spek to a kyng snce lordes or coun=
sellers

⁋ A trine Aspect of the mone to venus is
a good day to gite frendes and louers & to
make mariags

⁋ A trine aspect of the moone w marcu=
ri is a good day to resen and to mak acopt
and to mak bargens w marchauntes and
to put childzen to skole and to mete with
wise me and good to tak a Jurney

⁋ A quartil aspect of saturne w the mone
lett euert ma be ware that day a bout
what matter or bisines he geth a bowt.

⁋ A quartile aspect of Jubiter with the
mone is a good day to cosull with prelats
and iudges

⁋ A quartile aspect of Mars with the
moone is an euell day to sek ane fauer or
frendship of knyghts or squiers or of me
of warre

A quartile.

⁋ A Quartile aspect of ꝑ sone wt ẏ mone is an euell day to goo about ani gret matter

⁋ A Quartil aspect of venus wt ẏ mone is a good day to goo about ani thyng and speciall hit is a good day to mak a mareg or to gett loue &c

⁋ A Quartill aspect of Marcuri wt the mone is a good day to send forth ambassitors or messangers good to tak a Jurney good to bai or to sell or to mak marchauntes or to mak a reckning or acount

⁋ An opposicion of Saturne wt ẏ mone is an euell day to goo about eni gret matter

⁋ An opposicion of Jubetor wt ẏ moone is a good dai to mak sewt or seruice to ani gret man spirituall or temporall or to goo to him for ani other matter

⁋ An opposicion of the mone to Mars is an euel dai to tak seruants or to sek frendshep for hit is a day of malice anger and strife and war & euel for other matters

⁋ An opposicion of the mone to the sonne is an euill day to medell with gret mē or
with

with any gret matter.

¶ An opposicion of the Mone w' venus is a good day in all matters to be don.

¶ An opposicion of y' Mone w' Marcury is an indifferent day

¶ A coniunction of Saturn with y' men is an euill day to goo about any gret matter specialle to medell with old men or w' laborars or chorlish persons

¶ A coniunctiō of Iubiter with y' mone is a good day to make pece and to tretete for any matter how be hit verite & iustice wyll this day haue interest.

¶ A coniunction of Mars with y' Mone this is a day of anger of warre of stryfe and bate wherfore maryd men that hath shrewes to theyr wyues must not dysplese them for fere of after clappes.

¶ A coniunctyon of the sone w' the mone is an euyll day for all thynges. how be it it may be good to do that thyng that no man shall knowe

A cōiunctiō of y' mone w' vēus is a good day

to mak a mareges oʒ to get loue of womē
is so couerse and good to hire mē & womē
to saruice

⁋A coniunction of þ mon w̄ Marcuri is
good to bay & to sell & to make marchādis
to take a Iurney &c

⁋Whan ther is no aspect nyght to the
mone oʒ far from an euell coniunctiõ þ
day is good foʒ all maner of thyngs

⁋The mone coniunct in the dʒogons hed
is veri good And coniunct with the taile
of þ dʒagon is euell

⁋The mone Iunct w̄ good steres is good
And Iunct w̄ euell sterres is euell

⁋The .xi. capitle doth shew of flenbo-
thomi: oʒ letting of blod

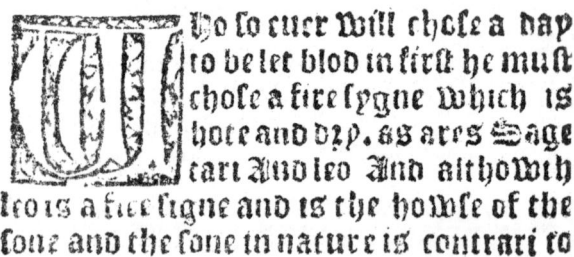

Ho so euer will chose a day
to be let blod in first he must
chose a fire sygne which is
hote and dʒy. as ares Sage-
tari And leo And altho with
leo is a fire signe and is the howse of the
sone and the sone in nature is contrari to
the nature of the moone foʒ þ sone is hote

and drye the moone is colde and therfor ye howse of the sone whiche is. Leo in hym is no good lettyng of blod. Secundaryly he must choose an Ayeryssh sygne which is hote and moyst As the fyrst medieate of Libra. And the wole sygne of Aqury. and althowgh ye Gemini and ye second mediate of Libra be Ayerysh. yet the Galax hath his course that way in ye which be many sterres of the nature of Mars. Wherfor ye mooone beyng in that place is combustid And therfore in that sygnes at that tyme is not good to let blod Therdli watterissh sygnes which be moyste and cold be good to let blod As Cancer & pisces, And Scorpio althowght of nature he be waterysh yet not withstondyng he is an oppositiue sygne of the exaltacyon of the Moone for Taurus is the exaltacyon of the moone & Scorpio is oppositiue in descendyng of ye mone Also Scorpio is the howse of mars Wherfor exhaustyng of blod in this thig is not good Also ye coniunctyon of ye mone w the sone. iii. dayes before and. iii. dates

D.i.   after

after is not good lettyn of blod. Nor ẏ coniunctyon of the mone with sature. Nor ẏ coniunctyon of the mone with mars and iz. degres before an as many after. Nor the oppsicyon of the mone with the sone. Nor the oppsycon of ẏ mone with Mars is not good to exhaust blod. Also a quatile aspect of the mone to the sone. And a qaurtile aspect of the moone to saturne. And a quartile aspect of the moone to Mars. all thes aspectes be not good to minishe blod for the Moone beyng in thes Planetes & Aspectes is contrary to the nature of mā consynyng flebotomy

℞ A tryne aspect and a sextyple aspect is good to let blod so be hit the moone be in a good sygne ℞ Also the moone coniunct w̄ Jubyter in a good sygne is best lettynge of blod ℞ The moone coniunct to venus or Marcury althowh they be good cōiuctions yet for as muche as the mone is veri nyght the coniunctyon. of the sone hit is not good to let blod ℞ Also ẏ mone being in a good sygne hauyng a sextyle or a tryne

ne aspect to Jubiter oʒ benus. oʒ Macuri
is best to let blod foʒ thes aspectes be good
⁋ Also the mone coniunct with the heed
oʒ tayle of the dʒagon beyng with in .xii.
degrs befoʒ oʒ after is not good to exhaust
blod. ⁋ Foʒder moʒe no wayne noʒ mem
ber of man owght to be tochyd oʒ hurt w
any insterment oʒ ferne whan the mon
is in any sygne that hath domynyon of ẏ
member . wherfoʒ in all maner of exhau
styng of blod from man oʒ woman ẏ day
and the tyme must be electyd ⁊ poyntyd
by som doctoʒ of phisick oʒ Astronamer.
what vayne , shold be oppnyd foʒ and yf
a man do lett blod of a vayne namyd cep
halica which is a vayne foʒ the hed ⁊ the
signe be in Aries. here ẏ flebothome doth
gret harme thorow his ignorance ⁊ thus
in lyek maner he may offend in lettynge
blod in any other place of man. ⁋ Foʒder
moʒe ẏ .iiii. fyʒe sygnes be good foʒ flemy
tyck parsones foʒ ẏ sygnes be of nature
contrary to the complexione therfoʒe Con
traria cōtrariis curāte. So in liek maner
fierish signes is good foʒ melancoly me

And watery ſh ſygnes or good for col-
rycke men And I ſay now a dayes
to vſe lettyng of blod except for
the ſquynce þ peſtylence or for
the kyndes of lepery, is not
good, for no mā can exhauſt
euyll blod but he muſt tak
good blod with all. And
þ lyf of man is in þ
blod for þ ſpryt
of man is in þ
blod. And
ther be
other
wayes to clenſe or to puryfy the
blod wout flebothomy or
lettyng of blod

¶ The .xii. Capitle doth ſhew, how
whā & what tyme a phiſicyō
ſhold mynyſter medycynes

✿ ❧ ✿ ❧ ✿ ❧ . ✿ ❧ ✿ ❧ ✿ ❧

hypocrates

Ypocrates saith ỹ phisicion is a blynde phisicion ỹ knoweth not Astronamye for Astronamy doth shew how whā and what tyme al maner of medsons owght to be ministred wherfor I do say yf an expulcyue medycyne be takyn inferpally. And the moone beyng in. Ariete. Tauro. or. Capricorne whiche takyth theyr name of terrestyall beestes namyd. The Rame the Bull and the Goote which hath doble belys, one is molifycatyue chewyng theyr cud ỹ other is digestiue mow by reson of the influence of the celestyal sygnes of lyek properte. shal cause the syck parson other to vomyte or els shall cause nauseons or eructuatiõ. or cause the substance that is in the stomacke to ascend to the orifice of the stomacũ and this doth lett ỹ operacione of suche medicynes specyally yf Saturne by a quartil Coniuntion or an opposicyon dothe geue theyr fortytudes to ỹ sayd sygnes. Wherfor let phisicyons be circumspect in thes mater

matteres and in all other. for hipocrates sayth in his fyrst pronosticates. Ther be sertyn celestyall thyng in the whiche the phisicyon must ͵puyd and he must be of gret ͵pudence. ꝛc. Wherfor Anice saith of celestyal & terrestyal matters happyth. ꝛc. Wherfor hit is to conclud that euery phisc̃on shold know ẏ nature & conplexiõ of ẏ sygnes and ẏ planetes. w̃ ẏ aspectes of them and to haue in a memory the cretyck dayes. And not to forgett ẏ operaciõ of the sygnes which be attractyue, retentyue dygestyue and expulcyue ☙ Wherfore I do say that the vertu of actraction consistith in hete and drynes And he that wyll confort and help ẏ vertu of actraciõ must do hit whan the moone is in a hote sygne as in Sagittary ad Leo. ☙ To cõforte or to helpe the vertu of retencyon is whan the mone is in a cold signe & dry as in tauro and virgo ☙ The sucur & helpe the vertu of dygestyon is whan the mon is in a hote signe & a moist. as in Gemini Libra and Aquary. ☙ To helpe expulcy
ues

ues is whan the moone is in a cold and a
moyst sygne as Cancer Scorpio and
pisces. ☞ Whan Uenus regnyth
in a good aspect to purge Coler.
Whā the sone doth regne in
a good aspect, purge flem
me whā Jubyter doth
regne in a good aspe.
ct, purge Melan
coly, ☞ Whā
ther is a cō
iunctyon
or a
quartyle aspect of saturne with the
mone minister thā no medycyne
uor whā Mars is in a con-
iunctyon. &c.

¶ The .xiii. capytle doth shew of
sowyng of sedes. of settyng
of herbes. and
plantyng
and
Greffyng of
trees.

He moone beynge in Tauro is
good to sow sedes to set herbes
to plant and graff trees takig
the tyme of the yere. beyng in
a sertile or a tryne aspect

⁋ The mone beyng in Cancer is good to
sow sedes althowh it be a movable signe
And so hit is in virgo haupnge a sertile or
a tryne aspect w to Saturne & ptholomy
sayth in the fyrst of his quadriptites hit
is good to plante

⁋ The mone beyng in libra hauig a good
Aspect is good to delue & dygg & to goue
to plow and to ere and to sow And so is ÿ
moone whan she is in caprycorne

⁋ The mone being in libra is goed to set
herbes & to plant tres haueg good aspect

⁋ Hereis to be notid that all thes thinges
is not good to goo about if Saturne or
Mars haue ani euell aspects

New

¶ Now to conclud I desier euere mā to tak this lytil wark for a past time. for I dyd wrett & make this bok. in. iiii dayes and wretten with one old pene with out mendyng and wher I do wret ẏ sygnes in Aries in. Taurous & in Leo is for my purpose it stondyth best for our maternal tonge

¶ FINIS

¶ Enpryuted at London in ẏ Fletestrete at the sygne of the Rose garland by Robert Coplande.